A. R Dresser

Bromide enlarging and contact printing

How to do it

A. R Dresser

Bromide enlarging and contact printing
How to do it

ISBN/EAN: 9783742827326

Manufactured in Europe, USA, Canada, Australia, Japa

Cover: Foto ©Andreas Hilbeck / pixelio.de

Manufactured and distributed by brebook publishing software
(www.brebook.com)

A. R Dresser

Bromide enlarging and contact printing

ONTACT PRI

HOW TO DO

———❧———

BY THE AUTHOR (

PREFACE.

SINCE the publication of my little book upon "Lantern Slides," the publishers have urged me to produce a similar work upon Bromide Paper. I have done so, and this book is the result. Encouraged by the reception accorded to my first effort, I hope to meet with an equal success with this. The reader may be sure of this—the matter is original, and within my own experience and practice. He may therefore rely upon it that nothing is advised which is impracticable. Whether it is of value must be left to his kind judgment to say.

A. R. DRESSER.

BEXLEY HEATH,
1891.

Synopsis of Chapters.

The "TALMER."

The optical and mechanical properties of this unique

HAND CAMERA

place it ahead of all other instruments. Its advantages and special features cannot possibly be described in an advertisement, but will be found enumerated in a 16-page pamphlet, which will be forwarded free on application. The No. 3 instrument is fitted with

F/6 Euryscope Lens.

NEW APPARATUS LIST NOW READY.

$\frac{1}{4}$ pl., 5, -

14;6 for 3

$\frac{1}{2}$ pl., 9 3

27/- for 3

Dark Slides. Best Quality.

Fitted to any camera. Careful fit and register guaranteed. All sizes made. Slides to fit Lancaster's and other cameras in stock. On receipt of value and postage a slide will be sent on approval for inspection. Over 10,000 of these slides have been fitted and sold in two years.

TALBOT & EAMER
7, 9 & 11,
Exchange St.,
(Facing Town Hall)

Telegraphic Address:
"Eamer, Blackburn"

BLACKBURN.

Bromide Enlarging: How to Do it.

CHAPTER I.

INTRODUCTORY. A MILD PANEGYRIC OF THE PROCESS.

THE advantages which the Bromide process possesses over other methods are possibly still unknown to a great many amateurs; otherwise it is difficult to account for the tenacity with which they adhere to such an uncertain, tedious, and inartistic system of printing as the albumen-silver, or the somewhat expensive and complicated platinum process. I have worked bromide paper, both for contact printing and enlarging, for several years, and the experience I have gained with it convinces me that both for large and small work no modern process is its equal in any of the respects by which the value of a printing process is to be judged. It is by the light of that experience that I have undertaken to write this short pamphlet on the subject, and in the hope that my fellow amateurs, who have hitherto held aloof from this most beautiful process, will be induced to give it a trial. If they do this I am confident that, like myself, they will cast aside all other methods of printing in its favour.

Let me endeavour, briefly and clearly, to lay before my readers the points of superiority of bromide to other printing processes. In the first place, one is perfectly independent of daylight, since the exposures may be made to artificial light, and thus bromide printing can be done at any time—a distinct advantage to the amateur whose daily occupation denies him the opportunity of utilising the rays of the sun for printing purposes. Furthermore, as a constant source of light is employed, as soon as the correct exposure is ascertained for a given negative at a certain distance from the light, the danger of over or under exposure is removed, and any number of prints from the same negative may be made without any doubts as to their correct exposure. This simplifies matters and renders good results a practical certainty.

The greatest advantage of all, however, lies in the fact that hand camera workers are able to enlarge from their negatives at comparatively small trouble and expense, and the initial outlay for big cameras and lenses is avoided. The labour of carrying heavy and cumbersome apparatus is also a thing of the past. Indeed, I think I may safely say that the hand camera owes a great deal of its popularity to this very fact ; and I am also persuaded that small stand cameras—say the quarter-plate and the five-by-four— are more frequently employed now-a-days than they were a few years ago, owing to the facilities which the bromide process gives for dealing with the negatives.

There is, I know, a feeling with many, that the development of bromide papers is an operation of parti- cular difficulty, but I hope in the course of the following pages to prove that this is not the case ; I will, however, add that it is governed very considerably by the expo- sure and the quality of the negative, and that it can be much simplified in several respects by the employment of novel methods of working hydroquinone and eikonogen, which I shall describe. If the first is correct—or approxi- mately so—and the latter suitable, then it is no more difficult to develop a bromide print than it is to print out an albumen-silver picture in the printing frame.

I also hope to show that there is considerable command both over the nature and quality of the resulting pictures. There are also certain advantages which the Bromide pro- cess has during development, such as the ease with which local doctoring of the print may be accomplished, and these advantages do not obtain with any other printing process with which I am acquainted.

In addition to the points which have been already put forward in praise of bromide paper, I may be allowed to mention that there is a choice of papers—white or tinted, rough-surfaced, grained, or smooth—with which no other process can pretend to vie. Its possibilities of tone (in the strict sense of the word), color of image, and brilliancy or softness at will, indicate the process as one combining a greater number of good and artistic potentialities than any extant photographic method. Nor is this all that can be written in its favour. The great sensitiveness of bromide paper to light and its power of responding to modifications of the developer stamp it as the most useful process of the day.

Like other good things, however, its virtues have been abused to its discredit, and negatives unfit for any other printing process (or even for the dustbin!) are sometimes "tried by bromide" as a last resource. This is a misfortune for the process, for the results, whilst they may be the best that can be obtained from that particular negative, are often judged by persons who ought to be better informed, as examples of the process itself. I do not think, however, that I need be afraid to assert that prints equal in every respect to the best that can be produced by any other means (platinum included) can be made with bromide paper. And this is supposing a good or first-class negative. Should this be not of the highest, then our method stands a full head and shoulders above any of the crowd of its competitors.

The continued improvements in the manufacture of the paper itself, the use of such a paper for its basis as artists prefer for *their* work, and new methods of dealing with the color of the image open out the possibilities of a largely increased usefulness for our already popular process.

I have endeavoured not to say too much in favour of bromide paper, but I could hardly have said less.

Having thus laid before my readers some of the special advantages which the bromide process enjoys, I will proceed to give a practical explanation of the several methods of working it, in the belief that if they will follow my instructions intelligently and carefully, soon will they be as enthusiastic in its favour as I myself am.

CHAPTER II.

IN beginning bromide work, one of the first things to be settled is the choice of a paper. The best advice I can give the beginner is to select a good make and always to use that, and that alone. A common mistake into which beginners fall is of flitting from brand to brand. This is a sure source of failure and unsatisfactory results, inasmuch as the properties and rapidities of no two commercial makes of bromide paper are identical. Like gelatine dry plates, each has its peculiar qualities, and the sequel to constantly changing one's make of plates is well known. Therefore, however trite the advice may seem, I repeat—keep to one make of paper. If it will guide the beginner in his choice, I may say that I have for years worked Fry's bromide paper and *still do so.*

This paper is sent out in three grades, namely, *smooth thin* paper, marked A; *smooth heavy* paper, called B; and *rough heavy* paper, or C brand.

The smooth thin paper is well adapted for printing from negatives having a great degree of fineness of detail; the smooth heavy paper is suitable for enlargements of portrait and landscape negatives which are to be mounted or framed; while the rough heavy paper is well suited to render the broader effects which result from the use of a paper such as obtains for drawing or water color purposes, or one in which the detail is partly subordinated to the effects obtained by an uncalendered paper. It will be found very suitable for landscape pictures of the character recently popularised by Emerson and others. These hints are not, however, intended in any way to fetter the photographer in his actions as regards choice of a paper. Often a decision can only be arrived at after prints upon two or more different grades of paper have been made.

But there is a fourth kind of paper which the amateur will find excellent for certain subjects—I allude to the "Roughest," which Messrs. Fry introduced at my suggestion. This was at first named by them "Naturalistic," and although afterwards that designation was abandoned, still the first name is occasionally used, but I protest against it as an objectionable one. This has not only a rough surface, but is toned, so that the "whites" of the picture are represented by the tone of the paper. Other developers than ferrous oxalate are not only possible, but advantageous with it. The paper is suitable for all work larger than half-plate, but is especially valuable for big pictures and subjects to be hereafter described.

The "Roughest" paper should be used for bold subjects, which depend upon effect rather than minuteness of detail for their pictorial success. For certain classes of landscape work, such as Emerson, Davison, Horsley Hinton, and others have made popular, it will give the required effect of "diffusion of focus" at will, without the necessity of having negatives of a special character.

Before attempting to enlarge on bromide paper, I recommend the amateur to thoroughly master the details of contact printing. By so doing he will acquire an intimate practical acquaintance with the process, as well as a good idea of comparative exposures and development, which will be useful to him at more advanced stages.

With the details of development I shall deal in Chapter VII., but before passing on, the opportunity must be seized to explain somewhat fully the all-important question of exposure. I cannot underrate the necessity of mastering this point to the would-be successful bromide worker. Three factors must be considered in dealing with the subject. They are: 1. The rapidity of the paper; 2. The characteristics of his negative; and 3. The value of the light. The inter-relations of these three factors must determine the length of exposure, and unless they are fairly well understood, exposure becomes to a large extent little better than chance-work.

As regards the rapidity of the paper selected, I think that for all practical purposes it may be taken as constant —that is to say, it does not vary to any appreciable degree one way or the other. This has been my experience of Fry's at any rate. Having once by trial, therefore, ascertained a proper exposure with any given negative, the judgment of the worker is called into operation to determine whether some other negative requires more exposure or less. This necessitates looking through the plate and mentally comparing its color and density with an arbitrary normal standard, for example, any negative with which the exposure *is* known. Passing on, now, to the second point, " the characteristics of the negative," it will probably be admitted that amateurs' negatives have an infinite variety of quality. I say *amateurs'* negatives advisedly, because a professional works—or *should* work—under such conditions as will enable him to secure a certain uniformity of character in his negatives, which is out of the question for the amateur, to whom such conditions are denied. It is for this reason then, that the amateur, as a rule, produces all sorts of results, and hence the importance of his studying every negative he proposes to use in bromide printing. Clearly, a " hard " negative will need a very different exposure to one which is full of half-tone, as, indeed, will a dense image need proportionately more exposure than a thin one. Again, the actual color of the image must be taken into account. A negative of a yellow color, due to development with pyrogallic acid in which no sulphite of soda has been employed, or one stained yellow by hydrokinone, requires more—considerably more—exposure than a clear delicate black image. A negative which is green or red-fogged—not so frequent an occurrence now as in earlier days of gelatine plates—although it may give a capital print, yet will it require a very long exposure, and as a rule intensified negatives do not make quick printers. Sometimes it may be advisable to alter the color of a negative for this very reason, and a method of getting rid of green fog upon negatives will be found in Chapter IX. If the negative be too thin it may be intensified, or, on the other hand, if too dense it should he reduced. This will give a little trouble and may cause delay just as the reader—always in a hurry to begin and "do something"—is hoping to make a commencement; but should the best results be desired it is preferable and necessary to make the best of one's *negative* before endeavouring to make the best print from it. This is only common sense and I leave it with him. There are negatives that are so violent in their contrasts, or so lacking in that respect as to be incapable of modification by chemical means. For these there only remains—a forlorn hope—treatment by varying the actinic strength and color of the source of light from which the exposure is to be made, and this brings us to the discussion of the last factor of the three which have to be considered in the matter of exposure.

It is, perhaps, too much to describe the alteration of the actinic strength of the light as a "forlorn hope," for within certain limits the character and strength of the light to which the exposure is made may considerably affect not only the time of exposure but the resulting print itself. The reason of this is, that the stronger (or richer in actinic rays) the light, the greater the power of penetrating the denser deposits of the negative and *vice versa*. I shall not here pursue this subject, because in dealing with the practical side of contact printing in the next chapter, full explanatory details will be given and points of practice touched upon.

The kind of negative that is most suitable for reproduction upon bromide paper is one that stops short of thinness, but *is full of detail* in the shadows. The abomination of the bromide enlarger is a negative in which the shadows are represented by clear glass, while the high lights are absolutely black. It is notorious that negatives which would not yield good enlargements in carbon or platinum, may often be turned to good advantage in bromide work. This is undoubtedly a "feather in the cap" of the process, but is also to some extent its greatest bane, inasmuch as it fosters the idea, that almost any kind of negative will do for this purpose. To my knowledge professional enlargers constantly have negatives sent to them which, strictly speaking, are unsuitable for enlarging purposes; although, of course, some kind of result has to

be obtained from them by hook or by crook. Clear
negatives, not too dense, with a good scale of gradation,
and plenty of detail in the shadows, such as are most
suitable for bromide work, are, I am afraid, none too
common ; but I hope, for the reader's own sake, and the
process's, he will always endeavour to get them—and
succeed.

CHAPTER III.

CONTACT PRINTING. THE LIGHT, AND HOW TO MANAGE IT. EXPOSURE. IN THEORY AND IN PRACTICE. THE NEGATIVE: HOW TO PREPARE IT FOR CONTACT PRINTING. PRINTING FRAMES. MASKS AND GUIDE POINTS. THE PAPER. OPALS AND IVORY FILMS.

TO make a contact print the only piece of apparatus required is an ordinary printing frame. It is necessary that this should be two sizes larger than the selected negative. By this is meant that if a half-plate ($6\frac{1}{2} \times 4\frac{3}{4}$ inches) is to be printed from, a frame of 10×8 inches inside measurement will be a convenient size to work with. The reason for this is, that some space for moving the negative is often required, and also that perhaps the most attractive and artistic way of producing a print is to have an inch or two of white margin round the portrait or view. This is the method generally adopted by engravers, and one that has stood the test of time. If the print is to be trimmed for mounting upon cardboard, or is to be placed under a cut-out mount, then no margin will be required ; but even in that case it is an advantage to be able to adjust the position of the negative in the printing frame, and obviously this cannot be done, unless it is of larger dimensions than the plate which is inside it.

The frame should be fitted with a piece of flat glass free from air bells and defects. It is upon this piece of glass that one's negative must be placed. Here let me refer to one small matter which should always receive careful attention. Clean and polish both the front glass of the printing frame and the back of the negative *in every instance.* Should this be neglected, whatever dirt or markings may

be thereupon will be reproduced in the print or enlargement. This is one of the little things which go towards
making the difference between good and bad work. Another
point not to be overlooked is the use of the brush for
working out pinholes, scratches, and other defects from the
negative. A good sable brush is necessary, for, unless
there be a *point* to the brush, the defect is made worse.
For color use Indian ink, or, better still, a special pigment
made for the purpose and known as GIHON's opaque. A
few minutes spent upon the negative will well repay one's
labor. It must be borne in mind that pinholes and like
defects, if not filled up with color, will print as black spots
or lines upon the final picture. If carefully "spotted"
out, as the phrase goes, should the marks show as
white spots upon the print, they may be removed with
a brush and color to match that of the image, or a black
lead pencil brought to a fine point.

I have now brought the reader to the point at which he
has his printing frame and negative ready to begin work.
There is yet one other point to which I must refer. Most
negatives are a little ragged or damaged at the edges, and
this will spoil the appearance of the print, unless it is
meant for placing under a cut-out mount. In many cases
this is not intended, and a print "complete in itself" is
desired. To effect this a "mask" must be prepared, and
I recommend everyone to do this for himself. Nearly every
negative will require a special shape or size of mask. These
masks—which the reader will understand to be pieces of
paper of a non-actinic color, and the central portion of
which is cut out to allow the negative to be seen or printed
through—must be cut so that only that portion of the
negative which seems to form the ideal-shaped picture is
visible. Any paper, which will not allow light to pass
through and fog the sensitive material, will answer
the purpose, but nothing is better than the yellow or
orange paper which is often used to pack photographic
sensitive materials in, and which can be obtained from
almost any dealer. Having cut the mask of such a shape
as to suit the requirements of the negative, and with a sufficient margin to protect the bromide paper to the required

width all round the negative, the frame should be taken in the left hand, and the negative placed film upwards in the centre of the glass front. If several prints are to be made from the negative it may be, with advantage, fastened into position with a strip of gummed paper or stamp edging. The mask is then to be placed upon the negative in such a position as the reader's judgment dictates, and he will probably find it a saving of irritation, and the only means of ensuring regularity of result, to fasten that down in the same manner as the negative. Now we are ready to place and adjust the bromide paper, unless, again supposing several prints are required, it is thought advisable to mark *upon the mask* the position in which the bromide paper must be placed to secure the image in the centre of, and square with the sides of the paper. This will necessarily

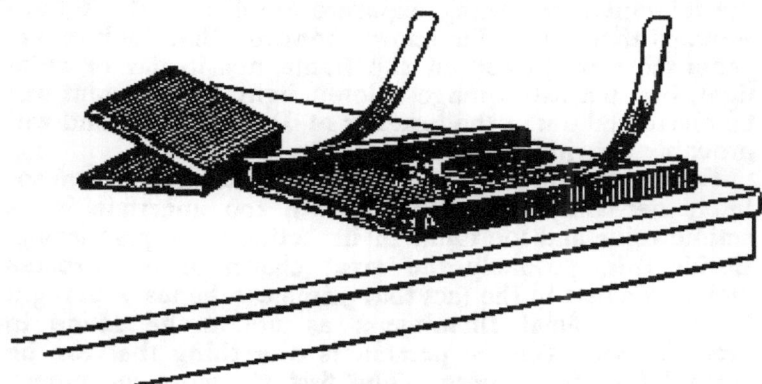

A.—The Printing Frame.
B.—The Negative.
C.—The Mask.

depend upon how much larger the bromide paper is to be than the picture itself. One printed from a half-plate negative, masked in the manner described, upon a piece of paper either $8\frac{1}{2} \times 6\frac{1}{2}$ or 10×8 has a sufficiency of white margin, looks very well, and is in a presentable form for examination by one's friends, whilst it is available for mounting or framing at pleasure.

The next step, then, is to place the bromide paper into

position upon the negative, the film or sensitive side of the paper towards the film of the negative. There are several ways of deciding which *is* the film side of the paper. The coated side always has a tendency to curl inwards. If this test is not sufficiently obvious apply the extreme tip of the tongue to the paper; the *absorbent gelatinous coating* will be instantly detected by its tendency to adhere to the tongue.

The paper being placed in position upon the negative, and a pad of a few thicknesses of dry, clean blotting paper upon the bromide paper, the back of the printing frame is placed over all and fastened up, and is now ready for exposure.

I must now for a short time leave the bromide paper in the frame awaiting exposure, whilst the source of light and the principles governing exposure are discussed. I have assumed that the reader knows, however, that the bromide paper must be placed in the frame, not in day or white light, but in a safe orange-colored light. This point will be elucidated under the heading of Development, and will probably be within his knowledge.

For contact printing a source of artificial light is absolutely necessary. Daylight is both too uncertain in its actinic value and too rapid in its action to be practicable. Beside this, probably the great charm of our process mainly consists in the fact that prints can be made at night by such artificial illuminants as are to be found in every house. Gas or paraffin is everything that can be desired for the purpose. The first, the more convenient and regular, the latter equally good in every respect if the trimming of the lamp be carefully and properly done. There is nothing to choose between them from the photographer's point of view. In my book upon the making of lantern slides I have gone into the theory of exposure to a slight extent, and if the reader will refer to that manual he will obtain all the information that is necessary for his guidance ("Lantern Slides," 2nd edition, page 22 and onwards; 1st edition, page 17 and onwards) upon this point. After further use I am still inclined to think that the board and gas jet therein described is the best for the purpose,

and I therefore recommend the reader to adopt that method
in its entirety, which I repeat for his benefit, with such
slight modifications as are required by the different process.

For this purpose a board marked in feet and inches, and
fitted at one end with a self-lighting burner or paraffin lamp,
as indicated in the illustration, is very convenient. It

A.—Gas burner.
B.—Printing frame.
C.—Board marked off in divisions.

should be so arranged as to be capable of connection to a
gas-pipe by means of an india-rubber tube, or, if a paraffin
lamp be used, the place where the lamp is to stand should
be marked.   All that is necessary is to mark upon the negative the number of seconds exposure, and the distance of
the frame from the light. From these data the exposure
can easily be correctly repeated at any time.

With regard to the distance between the exposing frame
and the source of light, two factors have to be considered,
the actinic strength of the light itself, and the density (or
thickness), or light-resisting power of the negative.   An
image which is very dense requires the source of light to
be rich in actinic rays. For a thin negative the reverse
holds good.   It is more convenient to increase the
distance between the exposing frame and the light
than to reduce the frame in size and brightness.   Given a
flame sufficiently actinic, its action can be readily modified
by varying the intervening distance. A paraffin lamp with
an inch wick, or a gas burner, consuming five feet per
hour, is sufficiently actinic to penetrate a negative as dense
as is generally met with.   Occasionally it may be necessary
to burn a piece of magnesium ribbon to print through a
very thick or yellow negative, but this is not often.

imperative. The law which governs this point is as
follows :—"The intensity of illumination on a given surface
is *inversely* as the square of its distance from the source of
light." Plainly stated this means that if at a distance of
one foot from a gas jet or paraffin lamp an exposure of five
seconds be correct, should the distance be increased to
two feet, the corresponding exposure will be twenty
seconds, *i.e.*, the original five seconds multiplied by two
squared. If the distance be increased to three feet, the
exposure will be five seconds, (the original exposure)
multiplied by three *squared*, or, say, forty-five seconds.
Stated proportionally, it stands thus—As the original
distance is to the increased or decreased distance *squared*,
so is the original exposure to the length of time which it
is desired to ascertain. As every school boy knows, the
first two terms must be expressed in the same denomination
—both in feet or both in inches.

$$1 \; : \; 3^2 \; :: \; 5 \; : \; x$$

$$x = \frac{5 \times 3^2}{1} = 45 \text{ secs.}$$

Theory indicates that the exposures given in the foregoing examples are equivalent, but there is a difference
in practice. Light possesses less power of penetrating
the densest portions of the negative as the distance increases, and we are able to make use of this property to
modify the result obtained. "The less the distance
between the source of light and the exposing frame, the
less the contrast between the high lights and shadows in
the resulting transparencies, and *vice versa.*" The practice
necessarily is—expose thick or dense negatives close to
the source of light, say 12 inches ; very thin negatives at
some longer distance, say 36 to 48 inches. Between these
distances for negatives of medium density. This is a
question for the reader's judgment. Having indicated the
lines upon which he must work, a few experimental
exposures will give him the confidence of experience.
There is one other method of modifying the action of the
source of light, viz., by altering its colour. A blue light
has a tendency to decrease the contrast—a yellow light to

increase it. A piece of very pale blue glass and one of very light yellow to interpose as may be required between the light and the negative will therefore be an additional power, and are recommended accordingly. It will be clear that there is considerable power of modifying the character of print which may be obtained from any given plate. If the negative be very dense, by which I mean if the deposit is very thick, and obstructs a large amount of light, then the distance from the gas or lamp must be short. If, on the other hand, the negative be a thin and transparent one, the reverse of that just mentioned, then there must be a greater distance between the negative and the light. Should the negative be a hard one, by which the reader will understand that I mean one, in which the transitions from high light to deep shadow are abrupt, then not only must the distance from the light be short, but the length of exposure must be prolonged, so as to minimise the contrast in the resulting transparency. Of course, the reverse must obtain if there be but little contrast between the lights and shadows, that is if the negative be what is technically known as " soft," and in such a case the use of the lightly-tinted yellow glass will be found of great service. It must be observed that the use of a yellow glass very considerably increases the exposure.

Having now explained the leading principles upon which the exposure depends, I shall endeavour to make clear a simple method by which the correct time for any negative may be ascertained, and as I have already written, with *one* negative of known exposure, it is but a matter of comparison of negatives for the worker to form a tolerably correct notion of what time should be given to other negatives. Let him proceed as follows :—Take the frame as we left it on page 11 ready for exposure, and holding it at a certain distance (say 24 inches) from the gas, let him expose the whole negative for 5 seconds. At that moment he should place a piece of cardboard so as to shield a portion (say one-fifth) of the negative from the light, and give a further 5 seconds' exposure. Then let him move the card so as to increase the part protected from the actinic flame to two-fifths, and again make a five second exposure. And so on until the whole picture has received five different exposures, the first part or strip of which has received five seconds only, the next ten seconds, then 15 seconds, 20 seconds, and the last 25 seconds.

Sketch, showing series of three Progressive Exposures.

If this sheet be now developed, and, for the sake of an example, it is assumed that strip which has received 15 seconds be correctly exposed, the information sought for will have been at once obtained. But it may be that the best exposure may be somewhere between two of these given times, and the exposure can easily be estimated for that. Again, *all* the exposures may be too little, in which case a further experiment may be made giving the first exposure at say 40 seconds, and advancing by periods of 10 seconds at a time. Or, instead of five strips or exposures a larger number may be made. Having indicated the method to be followed, an intelligent reader will be able

to apply it in several ways; but for the sake of making
the matter very clear, I have added illustrations which will,
I think, leave nothing to be explained except the development. That is all dealt with in its own Chapter. In the
improbable case of total over-exposure in the example
given, the worker will, acting upon the information already
given him on the subject of exposure, either reduce the
time or make an exposure at a longer distance from the
gas flame.

Having made the trial exposure (for which a narrow strip
of paper will answer all the purposes of a full sheet), and
replaced the *trial piece of paper* with a fresh sheet, all
is now ready to make the actual exposure, which may
be done. This part of the work is then completed.
The procedure in development will be dealt with in a
separate chapter. This portion of the work is exactly
the same whether the exposure be by contact or through
the camera.

I shall also treat the question of dodging the exposure,
vignetting, etc., etc., in the chapter upon enlarging. All
the little wrinkles, upon which so much depends, are
practically the same for *either* method.

In concluding this chapter, I will add that, if the word
opal be substituted for bromide paper all through, the
same instructions will apply, there being no difference
whatever in the practical treatment of the two materials.*

*[We should like to add that all Mr. Dresser's Instructions for
Bromide Paper are applicable to our Ivory films.—*Fry Manufacturing Co.*]

CHAPTER IV.

COMING now to enlargements on bromide paper, I will first of all enumerate the various plans available, and then proceed to deal with them in detail. Enlarged prints can be made by making a negative to the size required from a small negative—a 10 x 8 from a ¼-plate for example—and then taking bromide prints by contact in the printing frame as described in the preceding chapters. For this purpose a camera the size of the negative required is necessary. The alternative to this is to make the enlargement direct upon the paper. This may be done either by projection, or in the camera, and the light used may be either natural or artificial.

As this little work is designed to treat of enlarging *direct upon bromide paper*, it would be foreign to its aim to enter into any description of the methods of making enlarged negatives. It is a long and troublesome process, and one which, from its comparative difficulties, is not likely to recommend itself to the every-day amateur.

I will therefore pass at once to the methods of enlarging more particularly adapted to bromide paper. These may be divided into two sections, first, when the source of illumination is ordinary day light, and secondly, when artificial light is used. The former is again conveniently divided into two subsections, viz., first, when a room is used as a large camera, and secondly, when a special form of enlarging camera is used. I shall treat these two methods and the artificial light one in three separate chapters, commencing with enlarging in the dark-room upon a screen. The easiest way to comprehend this method is to think of the ordinary process of taking a photograph as reversed. Imagine having set up one's camera, and focussed it before some beautiful scene. The negative is already in the dark slide, exposed, developed, fixed, washed, and dried. A light of wonderful power is placed behind the negative, (which is in natural colours)—and shining through it, is throwing an image—like a vast lantern picture—upon the space in front of the lens. In other words that the landscape in front is a *projected image* from the negative, and not as really obtains, that the negative is a microscopic reproduction of nature. The hypotheses just described is what takes place when an enlargement is to be made by the first method. A negative is placed in the carrier of an enlarging camera (see block page 21), light is directed through it, thence through a lens, and then upon a screen (see block page 25). The size

Reflector fastened to the wall outside enlarging room.

of the enlargement depends upon the distance of the screen from the negative, and the length of focus of the lens. I will proceed to describe the apparatus in detail.

First, as to the means of directing the light. The rays may be passed direct through the negative, by placing the camera at such an angle as to point the plate towards the sky. *It is generally better*, however, to use reflected light and fix the camera in a horizontal position. The accompanying sketch (see page 19) will show the position and arrangement of the reflector clearly. It may be of—and I give the materials in the order of cost—

Silvered glass, (looking glass)
A piece of opal,
White cardboard, (not yellow or tinted)
A sheet of tin or tinned iron,

or any material which reflects the light in a satisfactory manner. If the sky be clear and blue, the white card or opal will be best; but should the sky be cloudy—and it will not be unduly pessimistic to calculate upon this chance—the mirror will give the best results. Generally the preference may be given to the last mentioned.

The reflector must be adjusted at an angle of about 45° upon the outside wall of some room which is to be devoted to the purpose of an enlarging and developing room. If it can be permanently fixed so much the better. It will be always ready for use. If not, temporary arrangements can be made for fastening it outside a window or other opening. The ingenuity of the reader will come to his assistance in this matter, if he understands the purpose of the apparatus, which, in this case is to receive the vertical rays of light from the sky, and pass them *horizontally* through the negative and lens to the screen beyond. The reflector is fixed outside the room—all the rest of the apparatus is arranged inside.

Now pass from the outside to the room itself. All white light must be shut out, and the camera fixed or fitted up against the opening or window in a horizontal position. It should be firmly fastened in such a manner, that the bellows can be extended for purposes of focussing and adjustment, and also that while the exposure is being made there may be no vibration or movement. As to the camera, an ordinary landscape one may be used, if the ground glass screen be removed, and the negative to be enlarged be fitted and made fast in the dark slide, so that the light can pass through without obstruction. I give, however, a block and description of an useful enlarging

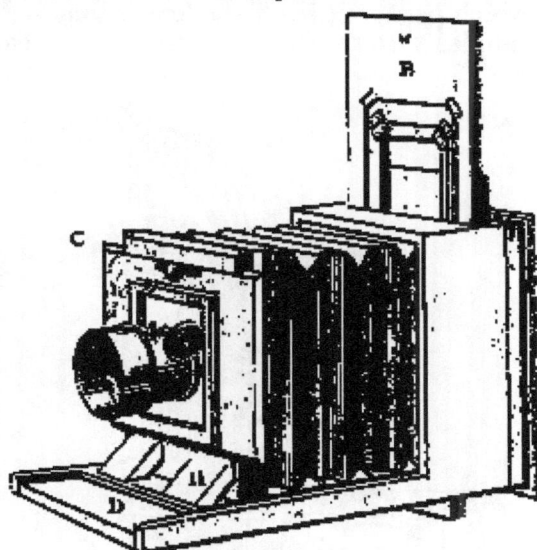

An Enlarging Camera.

camera. This will effect two purposes, and, at the same time, by showing the reader such points as are necessary, enable him to modify or alter his own camera if he so prefers. In the illustration on page 22 it will be seen that the enlarging camera is fastened to a wall by means of the rebated strips of wood A. At B is a nest of carriers with a vertical movement into which the negative is fastened. Here an advantage of the special apparatus is obvious, viz., a means of moving the negative *vertically*. By this device the centre of the negative may be made to coincide with the axis of the lens, and the best definition of the objective secured. *Horizontal* movement is obtained at C, which is a sliding front, allowing the lens to be shifted from side to side. By means of these two independent movements *any* position of the negative or the enlargement upon the screen is secured. D is the bottom

of the camera, of which no explanation need be offered.
E is the lens, which must be carefully and accurately fitted
upon the front of the camera, so that the axis—or imagi-
nary line which passes through the lenses longitudinally—
is at right angles with the back and front of the camera. I

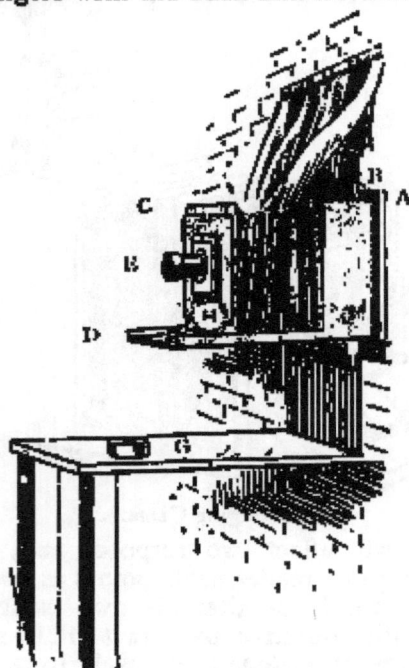

Enlarging Camera fixed to opening in wall.

may well state here, that the negative and the screen upon
which the enlarged image is to be projected must be in
parallel planes.   If this be otherwise, distortion of the
enlarged image will with certainty follow, and this is a
result to be carefully avoided.   It is also advisable—for
optical reasons—that the centre of the negative or that
portion of it from which an enlargement is required, should
be central with the axis of the lens.   In such a case a
larger working aperture to the lens is possible, and this
suggests another point, viz., that it is best to use as large

an aperture of lens as will give good enough definition. To use an unnecessarily small stop is a disadvantage; it prolongs exposure and tends to make the resulting pictures poor and flat in quality. At G is a special cap to fit the lens quite easily. It has a piece of yellow glass substituted for the opaque leather back. The object of this very useful adjunct is to enable the worker to see exactly where the enlarged image will fall upon his bromide paper, after it is fastened into place. Without such a device he must

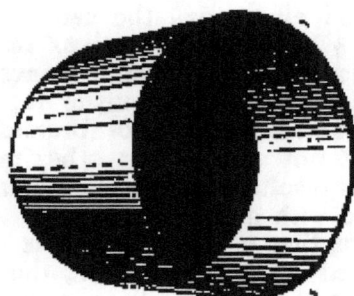

Special Cap, fitted with yellow glass.

work by measurement or guess work, and this is not well.

It will be seen from this description, that there is not anything very special about an enlarging camera that cannot be easily "rigged up" with an ordinary instrument. Obviously it is an advantage to possess an apparatus which needs no adaptation, and which has special mechanical arrangements to fit it for its purpose. Upon the question of the lens for enlarging, there is nothing better than a rectilinear lens of slightly longer focus than the longest side of the negative from which the picture is to be made. For a 6-inch negative an 8-inch lens will be satisfactory, and should cover well. The size of the enlargement does not affect this question in any degree, and to make this clear, let the reader once again think of the hypothetical case of the landscape being produced from the negative, and not *vice versa*. We can photograph a picture or view of any size with any lens, the only question which affects the definition of the image is the dimensions of the negative upon which the reproduction is to be made.

Let us now pass to the screen upon which the actual enlargement is to be effected. This may be a board, or any expedient for—in the first instance—focussing the picture upon, and afterwards fastening the bromide paper to. It should be borne in mind, however, that it must be always parallel to the negative, as before described. This purpose is most easily effected by having a screen to move in runners or grooves. To one further point must attention be given, viz., the centre of the screen or easel *must be in a line with the centre of the lens and of the negative*, for reasons already given. I have now described all the apparatus which is requisite for this, the most perfect and elaborate method of bromide enlargement. The reader must not assume that other methods will not produce as good results—but by this arrangement results can be produced with greater facility.

The actual routine of the process is as follows :—A negative is placed in the carrier at B ; the film side of the negative towards the lens, or the picture will be reversed in position—that is, a man's left arm will appear as his right. The picture is now brought to the right size upon the screen by moving it nearer to make the image smaller, or farther away to increase the dimensions, at the same time adjusting the position of the lens by means of the sliding front H, so that the image is sharp. It is a simple way, if unscientific, to move the screen about until the right distance is ascertained by experiment, and in practice the position is very quickly arrived at. In a later Chapter will be found full instructions for ascertaining the precise distance for any amplification, not only of the screen from the lens, but also of the lens from the negative. The image having being focussed and made quite sharp by the use of such a diaphragm as is found necessary, the yellow cap may be placed upon the lens, and a sheet of bromide paper fastened upon the screen. This can be most easily effected by having the face of the screen covered with a sheet of cardboard, which is excellent for focussing upon. It is also soft enough to fasten the sensitive bromide paper to by means of ordinary pins. Care must be taken to fasten out the paper as flat as possible, so as to secure good definition.

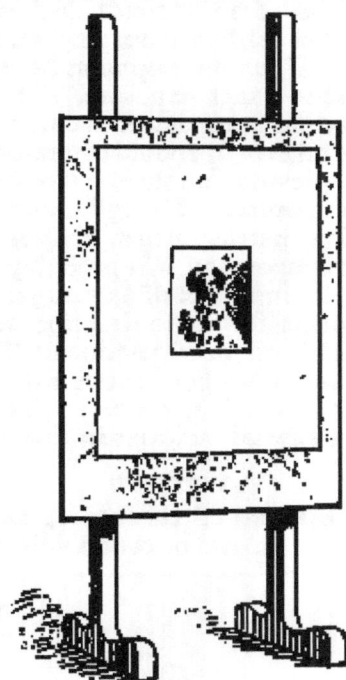

Easel or Screen, with Enlargement in position.

The yellow cap upon the lens will enable *the exact position the paper ought to occupy* to be decided upon, and if, after it is fastened into place, the picture be too high, or low, or too far to the right or left, it can be easily adjusted by means of the negative holder B, or the sliding lens holder C. Now comes the crucial point—the exposure. I cannot do better than refer the reader to the method of trial exposure for contact work suggested in the earlier pages of the book. Let him repeat the process here, and he will soon be able to acquire sufficient experience to enable him to judge an exposure correctly. Let him make an exposure of the whole sheet for say 10 seconds, and their shield with a piece of cardboard strips of the image, giving successive exposures of 10 seconds each until the whole piece is exposed, but with five varying and progressively exposed portions. Upon development he will easily make up his mind as to the exact exposure. The enlarger who can make five good pictures out of six sheets of paper is a very successful worker. I should therefore counsel the beginner to always devote one sheet out of every half-dozen to strips for trial exposures. He will easily be able to tear the sheet into five narrow strips, that is one for each remaining sheet of paper, and will probably find himself in pocket by this preliminary act of extravagance.

There now remains for me to describe several methods of "dodging" and of manipulation. These must be explained to enable the worker to cope with, and overcome the every day difficulties of the process. This arrangement of apparatus offers special facilities in this connection.

VIGNETTING.

To effect this a sheet of cardboard, about 10 inches square, is required.   An oval or round hole, about one inch

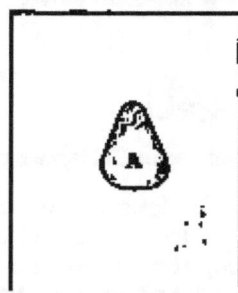

A. The hole   through which
the image   passes.

Cardboard Vignetting Mask.

or larger in diameter must be cut. The size and shape of the opening is dependent upon the picture in course of production—a round hole, generally speaking, will be suitable for a vignetted landscape, an oval one for a portrait. Mostly upon the amount of amplification of the picture the size of the hole depends. The vignetting is easily performed.

While the exposure is in progress hold the card between the lens and the screen, so that the opening is centred between the two. Keep it moving gently backwards and forwards during the exposure, so as to cut off all the outer portions of the picture which you don't want, and also to produce the soft shaded-off results which a properly vignetted picture should have. Hold the card nearer to the screen than to the lens at the start, and then gradually draw it away to proper distance, and having ascertained this keep it gently moving. A very little practice will enable one to modify the vignetting method to suit the composition of the picture.

Enlarging Apparatus, showing the method of using the Vignetted Mask.

The second use to which these vignetting cards are devoted is just as important as the first. When the picture is focussed on the screen, you will do well, besides studying what exposure to give it, to examine the image critically, with a view of deciding whether there are not parts of the picture which would not be improved by having a little more exposure than the others. Examine the high lights for details; if these, compared with the shadow, are not well translated, an auxiliary exposure will be of great service in bringing them out. Give the picture the exposure decided upon, and then take one of the cardboard masks, and holding it a few inches from the image, allow the light to act upon the shadow portion for a half minute or so longer (as you may estimate the increased exposure to be), taking care to keep the card on the move all the time.

Any part of the picture which may require it, can have an auxiliary exposure by proceeding in this way—in a portrait negative, for example, where the face has been well lighted and properly exposed, an ear, the back of the neck, the bust, the dress, or the coat (as the case may be) is possibly very much too dense, and the detail in those parts would not appear in the enlargement unless special means are adopted.

If enlarging from a landscape negative, and it is wished to cut out the sky, all that is necessary is to take a piece of cardboard cut to the shape of the foreground, and mask off the sky during the exposure, keeping the card moving so that no sharp lines are formed. Focus the image on the card before exposure, *roughly outline it with a pencil,* and so you will get it to the exact shape.

This very naturally brings me to the method of putting in clouds, which will be found simple and easy in bromide work. Focus and expose the negative from which an enlargement with a printed-in sky is required, taking care to shield or mask out the sky portion, as indicated in the last paragraph. It will be obvious that some negatives will be more difficult to mask off than others, and will require considerable skill or ingenuity to get the best results. I will instance a landscape with the spire of a church standing well up into the sky. This will require carefully working round, to keep the sky white. Now develop the print to its proper density. Wash off the developer carefully, but do not fix the print or allow white light to get to it. Change the landscape negative for a suitable sky one. I draw the reader's attention to this point, viz., that the sky negative must have been taken with the light falling upon it, from the same direction as was the landscape. It must also be suitable in the artistic sense. Focus this upon the screen, so that the picture is of suitable size to occupy the sky-space upon the half-completed enlargement. Place the yellow cap upon the lens again. Now take the developed and washed print—all wet as it is —and guided by the image, which comes through the yellow cap, adjust it for position. When this is to your satisfaction, cover the landscape or foreground part with a protecting shield of cardboard, and expose the sky portion. This will always be a shorter exposure than that of the landscape itself, and the card shield must be moved all the time of exposure to prevent anything like a hard line showing. It will be found, however, *that no harm will arise if the sky portion overlaps the foreground*, because development must only take place where the image *is* required. In the case of a tree, it may generally be exposed right over it, if carefully developed. (It is sometimes necessary, for artistic reasons, to secure the reflections of the clouds when water forms part of the composition of the foreground, upside down upon the water. It can easily be done by this method, but involves a third exposure.) The second-exposed, or sky portion, must now be developed with the aid of a plug or sponge of cotton wool soaked in developer, *in such a manner as to affect that portion of the picture only*, and when development is judged to be complete, the print fixed in the usual manner. The risk of the developer running over the already developed landscape or foreground portion, must be obviated by placing the print upon the bottom of a dish, and holding the latter at an angle and in such a manner that the developer, which escapes from the wool sponge will trickle away from, and not over, the completed portion of the enlargement.

Many other little dodges will suggest themselves to the careful worker as he proceeds. Some portions of the

negative may be too thin, and thus allow too much light
to pass through ! In this case exposure may be prematurely
stopped, allowing the rest of the picture to go on.   In fact,

Developing the sky portion of the enlargement with a plug of
cotton wool.

all the dodges and suggestions which have been thrown
out under the heading of contact printing are available
here. If the negative be masked as there directed, the
enlargement will be made with a bold white margin, and
this is a specially attractive form for unmounted prints, or
if toning by uranium is intended. The same care in
removing defects from the negative before proceeding to
work from it is necessary.   With celluloid films it is often
obligatory to fasten them *between glass plates* to secure
them in position.

CHAPTER V.

METHODS OF ADAPTING ORDINARY APPARATUS FOR ENLARGING.

IN a previous chapter the method of enlarging where a room can be devoted to the purpose has been dealt with. It is not everyone who has the convenience for doing this. To others, the following suggestions as to the use of the ordinary camera and apparatus are offered. In the first place I should like to state that an ordinary camera is all that is *necessary* to make enlarge-

Placing the negative against a window.
B represents a large sheet of Brown Paper.

ments. With a very little ingenuity the reader may easily adapt his own camera, *if it is large enough*. Perhaps the

simplest way of all is to fasten the negative, from which
an enlargement is required, against the window pane, to
which has previously been attached by a piece of gum paper
a sheet of ground glass as large as the negative.   Suppose
that a whole plate enlargement is wanted from a quarter
plate negative.   To effect this a camera at least as large
as whole plate will be required.   It is best to take a large
sheet of brown paper, cut a hole in the middle of it just
the size of the negative, and place the paper upon the
window, and the negative upon it so that it occupies the
cut-out space.   This prevents any extraneous light from
entering the lens of what is now the enlarging camera.
If this be not done, there would be a danger of the enlargement being veiled or fogged, owing to the incursion of
diffused light through the lens.   Now place the camera

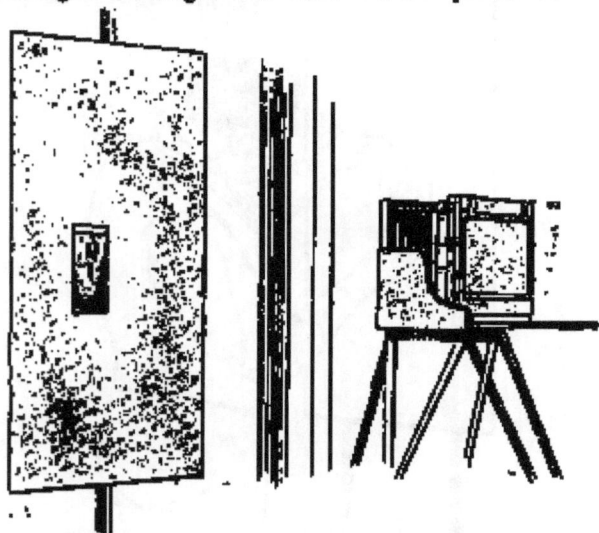

The Camera in position opposite the negative.

opposite to the negative.   Focus the latter in the usual
way, just as obtains in making a landscape or a copy negative from an oil painting.   That is to say, if the picture
upon the ground glass is not sufficiently large the camera
must be taken nearer to the window.   If, on the other

hand, it is too large, then must it be removed farther
away. This is a matter which a very little ingenuity on
the reader's part will enable him to cope with. The only
thing that he must bear in mind is, that he will require a
rather longer extension of the bellows of the camera for
this purpose, than for making ordinary negatives. It may
be necessary in some cases to have a special camera for
the purpose, or at least, to have the bellows of the camera
lengthened. Sometimes it is necessary to use a lens of
shorter focus. Generally speaking, in enlarging it is best
to use as short a focus lens as is reasonably possible. By
consulting the table of amplifications in a later chapter,
the length of camera required to make an enlargement
with a given lens can be ascertained, or supposing there is
a short limit to the extension of his apparatus, he will
be able to compute the focal length of lens necessary, if
the work must be done with the camera he already has.

I have indicated one method by which the ordinary
apparatus can be used for making enlargements, and I
trust that between my description and the illustrative
blocks, the method of effecting the purpose is sufficiently
clear. I will, therefore, without further description, give

Another make-shift arrangement.

another illustration, showing an alternative method without
the use of special apparatus, as a suggestion to the reader.

I should like to add that if any special apparatus is to be purchased, there is nothing to compete with that described in Chapter IV., but, short of that, the make-shifts here described are all that will be necessary.

To such as desire to make enlargements without any trouble of rigging up their own apparatus, there are many simple and cheap enlarging devices, the cost of which runs to a few shillings only. It is hardly ever necessary to buy a lens for the special purpose. Messrs. Fry sell a very capital apparatus for this purpose. This is an easy device for making enlarged pictures, as simple

Fixed Enlargement Camera.

in fact as making contact prints. It is a long box divided into two compartments; in the central partition is fixed a lens. At one end (c) a piece of bromide paper is adjusted in a carrier specially made for it, and in the other at (b) is placed the negative. This, of course, has to be done in a dark room. The exposure is made by bringing the apparatus out to the light, say a window, and exposing for fifteen or twenty seconds to daylight. At night the apparatus can be held opposite to a gas or paraffin lamp, and moved about so as to equalise and diffuse the light. The apparatus is made in various sizes, and is very cheap.

CHAPTER VI.

ENLARGING BY ARTIFICIAL LIGHT.

IN the previous chapters it has been assumed that the enlargements would be made by the aid of daylight. This precludes a great many people from making enlargements at all. For their benefit, therefore, I will describe the various methods by which artificial light may be used. The simplest, or at least the most usual method, is to employ the magic lantern. It is necessary in this case to have a condenser large enough to cover the whole of the negative. This limits the advantages and possibilities of the method very considerably, because a condenser to cover even a half plate to the extreme edges with precision is rather costly. Still, however, it is a method by which very good results have been obtained.

It will probably be sufficiently descriptive if I say that the negative to be enlarged from, must be placed between the condenser and the lens of a magic lantern. The image is thrown upon such a screen as has been already illustrated and described, and sharply focussed. The yellow glass cap is then placed over the lens, and a piece of bromide paper fastened upon the screen in such a position that the image falls upon it.

With an oil lamp the exposure will probably be from five to fifteen minutes, with an oxy-hydrogen jet from one to five minutes, according to the density of the negative.

With small negatives this method is fairly good; it has, however, one great defect, and that is, that every speck or imperfection upon the film is much magnified upon the enlargement. This is due to the fact that all the rays of light are carried in *parallel lines* through the condenser, and that there is no *diffused light* used at all.

To remedy this, many methods have been suggested. I have occasionally made very good enlargements by fixing up the negative in any of the ways already described in Chapter V., and illuminating it by moving about behind it a piece of magnesium wire, or by flashing a magnesium lamp, or the circle of light from a magic lantern may be thrown upon it.   None of these methods, however, compare in the excellence of their results with the enlarging apparatus which I am now about to describe.

Its essential principle is the use of reflected light.   By this means all the good effects and qualities of high-class daylight enlargement are secured.   In one side of a square box is cut a hole into which is fastened a negative.   Inside the box, and at a distance of a few inches—say 8 or 9— from the negative, is placed a sheet of opal or very white cardboard.   This serves as a reflector.   In the corners of the box—that is, on each side of the negative—is placed a source of light, a paraffin lamp, an Argand burner, or oxy-hydrogen jet, according to circumstances.   The light shines

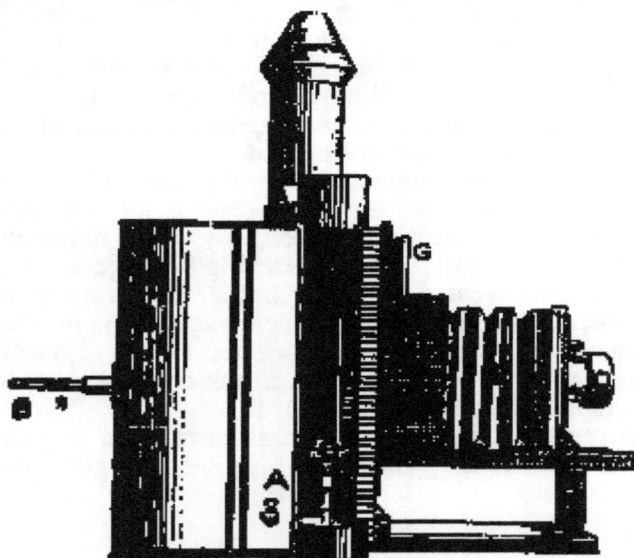

Artificial Light Enlarging Lantern.

directly upon the opal reflector, and back again from that through the negative.   By reflection the light is passed through the negative, mostly in straight lines, but with sufficient diffusion to imitate very closely indeed the effect of daylight. The more actinic the sources of light are, the more quickly can the enlargement be made. On the other side of the negative—that is, away from the reflector —the enlarging camera and easel is placed exactly as described in Chapter IV. In fact, if this box apparatus be substituted for the reflector which is fixed outside the wall, as in the diagram given in that chapter, the rest of the apparatus is the same. This method of enlarging is extremely simple, and I think it ought to supersede such methods as involve the use of condensers or other cumbrous devices for spreading the light evenly over the negative. That is the chief difficulty that has to be combated in working by artificial light.

It is a matter of considerable importance to decide upon the source of light, for it is not only a very tedious operation, but a great inconvenience, if the exposures occupy an inordinately long time. Taking all things into consideration, the oxy-hydrogen jet is perhaps the most easily obtained, and at the same time possesses sufficient brilliancy to make good enlargements quickly. Next to this a good four-wick paraffin lamp gives good results; but, lacking either of these, the use of a couple of good duplex lamps possesses but one disadvantage, viz., that the exposure is not short.

CHAPTER VII.

THE development of bromide prints and enlargements, if not more difficult than ordinary negative work, at least necessitates an amount of solicitude for the technical excellence of the results which may be absent from the other case. In making negatives, should the picture turn out too thin or too dense for printing purposes, it can be intensified or reduced to the proper degree; if it is stained, it can be cleared; if it is weak in parts only, it can be dodged in printing—in short, for most of the defects to which negatives are liable, a remedy is at hand.

This is only partly the case with developed pictures on bromide paper. If these turn out flat, hard, heavy in the shadows, veiled or stained, it is, as a rule, better to expose another sheet of paper; and, therefore, this knowledge should make the amateur attack the work with, if possible, more care than he devotes to negative development. For be it remembered, the negative is but the means to an end; but the print or enlargement *is the end itself.*

The development of bromide paper is conducted in the dark room, but, not being so sensitive as ordinary plates, an orange light, which is more pleasant to work by, should be substituted for ruby light. The same precautions for shielding the sensitive surface from actinic light must be carefully observed, or fogged pictures, with degraded high lights, will be obtained.

The utmost cleanliness must be observed, both in handling and developing bromide paper, more so, in fact, than in developing negatives, for of course the slightest mark or stain on the white parts—or, indeed, on any part—of the picture, will detract from its charm and purity. Bearing this in mind, the reader will do well to reserve a set of measures, trays, and dishes *specially for bromide work*, and always make a point of keeping them absolutely clean. If pyro development be habitually employed for negative work, the sink must be carefully cleaned before using it for bromide paper. Close attention to such matters as these, are among the conditions which help to produce success in working the bromide process.

A pledget of cotton wool is a most useful dark room adjunct, and a supply of it should be kept at hand for use as required. The enumeration of some of its uses will suffice to prove its value. With a pledget of cotton wool one can remove the excess of water from the film in the preliminary soaking before development; and dipped in the developer, the intensifier, the reducer, or the toning solution, it can be used either for local or general development, intensification, reduction, or toning.

There are three developers which are suitable for Bromide work—

> Iron,
> Eikonogen,
> Hydroquinone.

I have used each of them with good results.

Iron is the standard developer for the purpose, and has the fact that it is largely employed by professional workers to recommend its use to the amateur. The quality of picture it produces is probably unexcelled by other developers, but against these two points there are several disadvantages which the amateur would do well to weigh before deciding to adopt it for bromide work.

In the first place it can only be used *once* with complete success, whereas hydroquinone and eikonogen may be used more often. For under-exposed pictures, long experience has taught us that it is singularly ineffective. It is not by any means so easy or simple to prepare, does not keep well, and has a tendency to deposit iron in the image, which is a sure cause of yellow stains. Then again, it has the economic demerit of being a developer which is not used for any other kind of work.

In connection with the tendency to deposit iron in the

pores or film of the paper, it has become customary to
advise the use of an acid (preferably acetic) clearing bath
between development and fixation.   In fact, makers of
bromide paper, in the printed instructions which they send
out with each packet of their goods, advise that the print
should, after development with ferrous oxalate, be placed
directly into a clearing bath of weak acetic acid without
first being washed.   The object of this is to remove the
iron developer and prevent any deposit of the lime salts
which are always held in solution in tap water.   And this
purpose is effected very thoroughly, and at the same time
the color of the resulting print is thereby greatly improved.
It would be out of place in this manual, which is intended
to be a practical guide for workers and not didactic, to
labor at the theory of this part of the process, but it is my
duty to point out that there are dangers to be avoided at
this point.   It is absolutely imperative that if this acid bath
be used *before* the fixing operation is effected, *then must
great care be exercised in removing all trace of it before the
print is put into the hyposulphite bath.*   The danger
which has to be eluded is a simple chemical one.   It
is that, if any acid be carried into the hypo bath, sulphuration will take place.   By this is meant that the acid will
liberate sulphur from the hyposulphite of soda, and this
sulphur being set free *in the film itself* will, with certainty,
destroy the print in time.   The prints will go a yellow
color and the process will fall into disrepute.

It may be urged that if this be so the process must be a
poor one. But this is an argument that will not hold for one
moment. Every process requires to be worked under proper
and fixed conditions.   Platinum prints, for which great
permanency is claimed, will *and do* yellow in this very way
if the iron salt which is in them be not perfectly or sufficiently eliminated.   My advice, then, to the reader is not to
use the preliminary acid bath unless he is prepared to
exercise the most loving care in personally satisfying
himself that all trace of acid has been washed away before
fixing. If eikonogen or hydroquinone be selected as
the developer and to the fixing bath be added a proportion
of acid bisulphite of soda, all this labor of the acid

clearing bath and its thorough elimination will be avoided.

I will only add that much of my exhibition work has been developed with eikonogen, and I think that in any case the best results can be obtained with these developers, the instructions of bromide paper makers notwithstanding. If it be intended to tone with uranium, I advise the use of eikonogen (or hydroquinone) *in preference to* the iron developer. This, I am aware, is contrary to the advice of Mr. Weir Brown, who has done so much to make uranium toning practicable and popular, but further practice confirms me in my belief that for the average worker, either of these developers is preferable to iron *if toning by uranium is to be resorted to.*

To sum up, eikonogen and hydroquinone are not subject to the disadvantages of iron. One or the other of these developers is probably used by most amateurs for negative work. If supplemented by an acid fixing bath they give just as good tones as iron. Either developer can be used several times; both keep practically indefinitely, require less exposure than does iron, and by varying the proportions of the alkalies employed, the appearance of the resulting image may be considerably modified.

I have thus tried to place the advantages and disadvantages of the three developers fairly before the reader, and it will rest with him to decide which he will adopt in his work. Let me now endeavour to describe the manner in which I have employed them myself.

THE IRON DEVELOPER.

The formula which I have employed for the above developer is as follows:

Prepare an *acid saturated solution* of iron protosulphate by adding crystals of the salt to a jug partly full of warm or hot water. Say a temperature of about 120° Fahr. About 16 ounces of iron will be required to saturate each pint of water. Stir the liquor well with a wooden stick and add acetic acid, drop by drop, until a piece of blue litmus paper dipped into the solution just turns red. The solution must now be allowed to get cold and clear from sediment

or deposit, which may take 12 hours or more.   This
settling may be obviated by filtering the solution, but this
is a messy operation and one to be avoided by preparing
the solution 24 hours in advance of the time when it will
be required.

Also make an *acid saturated solution* of oxalate of potassium, proceeding in exactly the same way, but of course
substituting the potassium salt for the iron.   About
7 ounces of oxalate will be required to saturate each pint
of water, and remember that in making saturated solutions
there must always be an excess of the salt which crystallises
out when the solutions are cold.   This is not wasted.
Heat and more water may be applied, and the crystals
dissolved for a second making.   It is also well to keep the
vessels containing all saturated solutions in some warm
place, for the reason that if they get very *cold* the liquid
will contain less iron or potassium, as the case may be.   In
other words, there is another danger to be avoided with
the ferrous oxalate developer, viz., getting the solutions,
which should be saturated below normal strength.

Another, and a very convenient method of preparing
these solutions, consists in placing the chemicals in little
pieces or bags of muslin, and suspending them, with the aid
of a piece of string, from the top of the jug in which the
solutions are to be made.   This is a cleanly and correct way
of working, and it will be found possible to obtain a clear
solution more quickly by this contrivance than by the means
previously indicated.   The solution can be used to the last
drop by lifting out the bags.

In addition to the foregoing, a 10 per cent. solution of
bromide of potassium (the restrainer) will be required.
This is made by placing one ounce of the salt in a measure
and filling it up to the ten-ounce mark with water.

Suppose it is desired to develope a 12 by 10 enlargement that has been, we will assume, correctly exposed, take
6 oz. of the oxalate solution and add to it 1 oz. of iron
solution.   Remember that the iron should always be added
to the potash, and not *vice-versa*, otherwise a muddy solution
will form which will be inefficient for developing purposes.
To each fluid ounce of the mixed developer add one drop·

of the 10 per cent. bromide solution. This will constitute the normal iron developer.

Having placed the sheet of paper in the dish face upwards, cover it with cold water for a minute or two, drain and remove the excess of water with a clean pledget of cotton wool. Now flow over sufficient developer. This should be quite clear and of a reddish brown colour. The high lights should appear in about half a minute, then the shadows, and the whole picture be out in a few minutes. Inasmuch as the bottom of the dish should be of glass, the progress of development can be watched by transmitted as well as by reflected light.

To get an accurate notion of when full detail and density have been secured, the reader must form his judgment by looking *through* the print, and not at it. It is for this reason that the developing dishes are recommended to have glass bottoms, and for no other. Now is the time when the worker has his opportunity of deciding whether the range of tone, the depth of full shadow, the wealth of half tone are just such as he thinks the negative he is reproducing *ought* to give. My technical instructions may be good or bad, but he alone can be the judge whether the prints he is making from his negatives satisfy his own notions of what they should be, or his artistic aspirations.

Examine, then, the print carefully at this stage, and do not hurry on development to a close. With a suitable developing light there will be no fear of fog from that source. If undecided, wash off the developer so as to have time for fair judgment. Anyone can develop a print just as it (the print) chooses to come up; but the artist desires to leave the impress of his own feelings and choice upon his work. If the shadows, then, are going too deep, stop further development there. Are the half tones lacking in detail, or the high lights undeveloped blanks? Always have at hand when working a pledget of cotton wool or a *large* camel hair brush. Having washed off the developer from those parts of the picture where its action has gone far enough, locally develop the other and insufficiently-reduced parts with the brush and some fresh developer. By using a solution of potassium bromide in the same way as the developer for local development—namely, by painting it on with a brush—parts of the prints which are inclined to be too dense may be held back while the others gain in force.

I have now laid before the reader *all* the secret of manipulating his development. He will find full scope for his talent in every picture that he prints, if he desires to secure the best as the result. One word more, if the print be badly under or over-exposed, do not waste time and patience upon it, but *make another exposure*, and a more correct one.

When development is judged to be completed, the solution should be carefully washed from the back and the front of the print. Upon the whole the best method is to use a large handful of cotton wool, soaked in clean water, as a sponge. Pass this lightly over the surface of the print, to remove any deposit, and at the same time free the gelatine film from the developing agent.

It is now ready to be fixed in the following bath :

FIXING BATH.

Hyposulphite of soda - - -	4 oz.
Water - - - - -	20 ,,

In this bath, *which should be new*, let the print remain face downwards for 15 minutes, moving it occasionally to prevent air-bells from adhering to the film and so causing unfixed spots. It is imperative that the print should remain in this bath the full time of 15 minutes to ensure thorough fixation. The print being now completely fixed, the important operation of washing or removing the fixing solution from the film and paper begins. Let me advise the reader to proceed in one way, and that only. Success is certain and the washing process is speedy. It is the quickest and most reliable method. Lay the print face or film downwards upon a piece of glass, pass over the back lightly an indiarubber squeegee, an instrument which may be bought for a few pence from any dealer in photographic goods. This squeezes or presses out all the surplus liquid. Now place the print into some clean water for five minutes. Repeat the process ten times and washing is complete. Several prints may be washed at one time. Do not be sparing of trouble, as in this way more perfect

washing is secured in one hour than by any other method
in six times as long.

It may sometimes happen that there is a yellowness
all over the print. To remove that, after the washing is
complete, pass the print into the following clearing bath :

ACID CLEARING BATH.

Acetic acid	- - -	3 drams (180 drops)
Water	- - -	3 pints.

After this bath the print will require squeegeeing (as before
described) twice.

The alternative system of clearing the print before fixation is as follows. As soon as development is finished
pour off the developer, and *without washing* apply the
clearing bath above mentioned. Before placing the print
in the fixing bath it is *absolutely imperative* that the very
last traces of acid shall be removed from it, otherwise the
decomposition of the hypo will inevitably produce a yellow-
ing—or sulphuration—of the print, as before described.

The necessity of an acid clearing solution is one of the
drawbacks of the iron process on account of the risk just
mentioned. I have described both ways of employing the
bath, and the reader must choose for himself the one
which commends itself to him.

If preferred, the acid clearing bath (page 47) may
be used in conjunction with the ferrous oxalate developer,
and so the use of an acid bath before fixing may be
avoided. The results are good, but I have not had much
experience with it, so only mention it here.

CHAPTER VIII.

WHEN one knows how to use it, eikonogen as a developer for bromide prints is quite as good as iron. With those who advocate a weak eikonogen developer I do not agree. I have recently given it a most exhaustive series of trials, and I find I get by far the best results by giving rather short exposures and using a strong developer. Here it is :—

EIKONOGEN DEVELOPER.

Eikonogen	-	-	-	1 oz.
Sodium sulphite	-	-	-	4 ,,
Potassium carbonate	-	-	-	1 ,,
Sodium carbonate or washing soda	-	-	1½ ,,
Water	-	-	-	30 ,,

Mix with boiling water and allow to stand until the solution is quite cold, when it will be ready for use. The addition of a restraining bromide is not necessary. If the picture be over-exposed it is only necessary to dilute the eikonogen with an equal bulk of water. Otherwise, I prefer it at full strength. This developer is inclined to give a greenish instead of a black image if bromide be used with it, but it does not stain the paper unless it is old, or is much under-exposed. Should a pink discolouration after fixing be apparent, a very weak solution of cyanide of potassium will remove it without damaging the print. A great point in favour of eikonogen, is that it works very clear and does not stain the fingers or the developing dishes. It does not absolutely require an acid fixing bath, but works at its best in conjunction with it.

I now pass on to hydrokinone as a developer for bromide

work. If one cannot speak so well for this agent as for eikonogen, still the introduction of the acid fixing bath makes it of greater value.

The following formula will be found serviceable and reliable in practice. It has worked well in my hands :—

HYDROKINONE.

A.

Hydrokinone -	-	-	-	150 grs.
Neutral sulphite soda	-	-	-	600 „
Bromide potassium	-	-	-	25 „
Water to make -	-	-	-	20 ozs.

B.

Carbonate potassium	-	-	-	2 ozs.
Carbonate soda -	-	-	-	2 „
Water to make -	-	-	-	20 „

For use take equal parts for normal exposure.

It is essential that this developer be used only in conjunction with an acid fixing bath, or success is uncertain. The acid fixing bath is made as follows :—

ACID FIXING BATH.

Hyposulphite of Soda	-	-	-	-	4 ozs.
Acid Bisulphite of Soda	-	-	-	-	1 oz.
Water	-	-	-	-	20 ozs.

Immediately after development, *and without washing*, the print should be placed in this solution, and when fixation is complete should have hypo squeegeed out of it, as before described, before placing it in the wash water.

All the details and the method of development with either hydroquinone or eikonogen are identical with those given in Chapter VII. for ferrous oxalate, and it is unnecessary to recapitulate them. All the processes are the same, and the washing after fixing must be conducted with equal care and thoroughness.

I have in the earlier pages of this book dealt with the advantages and weak points of the various developers. It remains for me to add, that, although there latter claimants are essentially more easy to manipulate, and in that sense may be described as suitable for beginners, yet I fancy that having once mastered their use, no sufficient advantage will be discovered to induce a change to ferrous oxalate. Photographers used to this last developer, may be reluc-

tant to alter their method for the newer re-agents; but the
amateur who begins with eikonogen or hydrokinone, will
hardly care to change to the iron developer.

The acid fixing bath, for which a formula is given above,
is the simplest form for use. I add, however, for those
who cannot obtain the acid bisulphite of soda easily, or
who wish to make use of neutral sulphite, the following
instructions for so doing :—

Sulphite of soda (ordinary neutral) - -	2 lbs.
Strong sulphuric acid - - - - -	2 ozs.
Water - - - - - - -	1 gall.

Mix the acid with a pint of the water, dissolve the sulphite
in the remainder, add the two liquids together. Use the
solution so made to dissolve the hyposulphite of soda, in
place of ordinary water.

CHAPTER IX.

ALTERATIONS OF DENSITY AND COLOR. REDUCTION. URANIUM TONING.

IT sometimes happens, however careful one may have been during the processes which have been described in the foregoing pages, that the print is not all that is desired, and it is not altogether convenient or possible to follow the best course in such cases, viz., to make another print. In such a case the worker may desire to resort to intensification or reduction, or to one of the little dodges which I shall explain to improve his picture or to remove some of its defects.

Intensification.—It is generally held that bromide pictures cannot be intensified. This is not my view, but the methods by which it can be satisfactorily effected are limited. I know of only one, and that is Mr. Chapman Jones's method of bleaching with either mercury or copper, followed by re-development. I do not profess to pose as an authority upon such matters, but as far as my acquaintance with that method is concerned, it is to be relied upon as permanent, and it is certainly a very practicable and facile method of adding to density. To effect the change proceed as follows. Soak the print in clean cold water until it is quite evenly wet all over. Here let me say that it is imperative to *ensure* success in any of the after processes detailed in this chapter, that the print to be treated must have been previously *sufficiently fixed, washed,* and *dried.* The risk of failure is doubled if the last point be neglected. The print may now be placed in a bath of bichloride of mercury. This I prefer to make up and keep in a strong or saturated solution, to which ten drops of hydrochloric acid should be added for every five ounces of solution. For the actual bleaching, however, I prefer to work with

as weak a solution as will effectually bleach the print, and
proceed thus. I take one ounce of the strong solution and
add to it twenty ounces of water. In this bath I soak the
print until it is bleached, or has lost color. Should this
not be effected it is necessary to add, a few drops at a time,
more of the strong solution until the bleaching has taken
place. For the benefit of the entirely inexperienced I
may mention that when a negative or positive, the image
of which is composed of silver, is placed in a solution of
this salt of mercury (chloride of copper has the same effect)
it gradually changes from a black or dark deposit to a
lighter one, often becoming difficult to see at all, and it is
this change which is termed "bleaching." The change is
effected with greater or less rapidity, according to the
strength of the mercury solution and the amount of deposit
of silver in the film. I prefer to effect the change quite
slowly, and should not, therefore, object if the reader were
to start by only using a dilution of one part of strong mercury solution to forty parts of water, instead of the proportion first given. To be clear upon this point, the dangers
to be avoided are irregular bleaching or staining of the print
and a tendency to block up the shadows and half tones of
the image, and the process is best carried out if the action
be commenced with a very weak solution, which can be increased in strength according to the worker's judgment. I
also very strongly urge that this solution should never be
used for several prints in succession. What the chemistry of
the matter is I do not know, but this is certain—he will
get better results, or rather will run less risks of failure, if
my suggestion be adopted. The print being bleached must
now be washed, and real thoroughness is essential, for if
the mercury be not removed from the film the next process
will be a failure. The intensification will be irregular and
in patches, and the print spoiled. I have already on p. 54
advised the reader how to wash the print. Let him repeat
the operation. When washed the bleached picture is to be
re-developed. This may be done with new or old developer, and it may be either iron, hydroquinone, or eikonogen,
and in daylight or not. It is immaterial. When re-developed (and progress is visible), wash the print and dry it.

The operation is now completed. One word more, never leave the mercury solution within the reach of children. It is a very virulent poison. I do not recommend, however, that it should be labelled POISON in a horrifying manner. All chemicals are more or less poisonous, and this happens to be a little more so. Mark the bottle "Bichloride of Mercury," and put it either out of ordinary reach, or, better still, *under lock and key*. It is almost as important *to the photographer* to keep his solutions from being meddled with, as that they should not be used for wrong purposes, although doubtless a coroner's jury would not endorse this view.

Reduction.—Reduction can only be effected to a slight extent and requires considerable care. The dangers are two. Destruction of the half tone and irregular color (the image is of different patches of color). To avoid the first the reducer must always be used very weak and its action must not be maintained too long. I imagine that upon a bromide print all reducing solutions act more vigorously upon the half tones than upon deep shadows. The former are more easily attacked by the solutions. Use the reducer, therefore, only so long as this effect is not visible.

The best reducer is the ferricyanide of potassium and hypo bath:

REDUCING BATH.

Hyposulphite of soda -	-	- 1 part
Water -	-	- 10 parts
Saturated Solution of ferricyanide of potassium -	-	3 drops to each oz. of the Reducing Bath.

It is a good plan to immerse the print (after the reducing solution has been rinsed off it) in an ordinary hypo bath for half a minute, for a large excess of hyposulphite of soda is required to dissolve the ferrocyanide of silver formed as the exchange product. The formula given has a sufficient excess, but the course suggested is a safe one. *Only use the reducing bath once.*

The mixed reducing bath will not keep, and its action depends upon the amount of ferricyanide (not ferrocyanide) of potassium which is present. To increase the reduction, therefore, increase the amount of ferricyanide. All my remarks upon intensification apply here.   The fixed, washed, and dried print must be evenly moistened and placed in the reducing bath.   Local reduction may be stimulated by the use of a camel-hair brush upon that part which is too black.   In this way heavy shadows may be made more transparent.   Careful washing necessarily follows the use of the reducer.   A much more powerful reducing solution is the following:

IODINE AND CYANIDE REDUCER.

Cyanide of potassium	- - -	- 20 grains
Water	- - - - -	- 1 ounce
Tincture of iodine	- - - -	A few minims

This solution, which may with advantage be much diluted, is used in the same way as the preceding one, but is much quicker in action and requires more care in use.   This solution is a very useful one to have at hand for removing black stains or marks upon the film.   In fact any deposit of silver may be entirely removed by it with great facility. Or a slight deposit of fog may be dissolved away, and an apparently spoiled print made serviceable or available for toning, &c., &c.

There are other methods of altering the color of bromide prints, such as the sulphocyanide gold toning bath, such as is recommended for aristotype papers.   Being of doubtful advantage, I merely mention, and enumerate the formulæ for the reader's benefit, if he cares to try them:

PURPLE RED TONING BATH.

Sulphocyanide of ammonium	-	- 15 grains
Hyposulphite of soda	-	- 240 grains
Water	- - - -	- 2 ounces
Chloride of gold	- - -	- 1 grain
Dissolve the gold in a very little water and add last.		

This bath must be prepared 12 hours before it is required for use, as it needs some hours to mature its toning powers.

BLUE BLACK TONING BATH.

Sulphocyanide of ammonium	-	- 10 grains
Carbonate of sodium	-	- 2 grains
Water	- - - -	- 4 ounces
Chloride of gold	- - -	- 1 grain.

As in the previous formula, the gold must be dissolved in a very small quantity of water and added last, and time given to the bath to mature.   In these two formulæ the addition of a strip of bromide or albumenized or aristotype paper helps to mature the toning power of the solution. The prints to be toned should be put into the bath dry.

A very good sepia tone can be given to bromide paper by the aid of the Platinotype Company's ready-made intensifier.   This solution, which is, as its name implies, an intensifier, is also a toner, and varies the color in the direction of warm brown.   The solution is, however, too strong to use undiluted, and about two or three times its own bulk of water should be added.   Allow the print to dry (after fixing and washing) before altering the color.

There now remains another and perhaps the most important method of altering the color of bromide prints, viz., the uranium toning process.   This old friend has been revived and made both popular and practical by Mr. Weir Brown, of Croydon, who deserves the thanks of all friends of bromide for his labors in this direction.   It is a toning *and* an intensification process, and allowance must therefore be made for a slight increase in contrast and depth. The best formula is Fry's modification.   Being in two solutions, it keeps indefinitely, and is therefore always ready for use :

URANIUM TONING BATH.

1. Nitrate of uranium	- - -	20 grains
Acetic acid	- - - -	½ ounce
Water	- - - -	10 ounces.
2. Ferricyanide of potassium	- -	20 grains
Acetic acid	- - - -	½ ounce
Water	- - - -	10 ounces,

To be mixed in equal parts for use.

The toning process with this bath is precisely similar to the others just enumerated.   The washed and dried print must be soaked in cold clean water, and then be placed in the mixed solutions until the desired color is obtained, when the print must be removed and washed until the whites of the print, which may have become stained a little yellow, are white again.   This process is particularly applicable to C grade of paper, and the Roughest or " Naturalististic " Brand.

Mr. S. Herbert Fry has suggested the following method of using this process, and, as he claims some advantages for it, the following description of the method (which is written specially for this book) is added:—

"In comparing the work and the remarks at Society meetings and elsewhere, upon the subject of Uranium toning, it has always struck me that the one difficulty in connection with it, was the more or less troublesome and persistent yellow stain of the toning solution itself. This may be easily avoided by the following method of using the chemicals, and it is specially valuable when a very thick or Rough surface paper is being used. Take the dry print. I prefer one for this purpose developed with Eikonogen or Hydro-Kinone, but there will be no difficulty in dealing with an ordinary iron developed print if it has been freed from iron. Soak it in cold water and place it *face upwards of course*, upon a sheet of glass, or other clean impervious material. Have the toning solution at hand in a saucer or open vessel. Also have a large handful of cotton-wool which has been soaked in clean cold water and squeezed out. With this rub down the bromide print (face or film upwards) into close contact with the glass, getting rid of all air bells. To do this it will be necessary to have the cotton-wool sufficiently wet (that is holding sufficient water) so that the surface of the print will not be abraided. The paper should now be down flat upon the glass. Wring out the cotton-wool sponge and soak it in the uranium toning solution, then pass it deftly and rapidly over the surface of the print. Be careful that every portion is covered, and if necessary use more toning solution. In any case use enough and keep the solution, by means of the wool sponge, moving over every part of the print until the tone is satisfactory. At this point quickly wring out the sponge and sop up all the remaining toning solution from the face of the print. This done, wring out the wool sponge in clean water and rub over the surface of the print, being careful again to cover every part. It will be found that, owing to the toning solution having only touched *one* side of the print, and that side being protected by a film of gelatine, there is practically no yellow stain. What there is can be easily removed by washing with the wool sponge, or by placing the print still upon the glass plate under a water tap or hose, and keeping it moving. There are two great advantages of this method. First: Very little after washing is required, there being no toning solution in the body of the paper. Second: The exact tone required can be obtained, the necessity of carrying the toning process further than is actually required, in order to allow for the reduction which takes place, if prolonged washing is necessary, is obviated."

Toning a Print.

I see that Mr. Weir Brown has since recommended the process of over-exposure and development, and afterwards followed by toning with uranium, followed by reduction with the ferro-cyanide reducer. This is a refinement of the process of which enough is not yet known for me to make a reliable statement. I will close this chapter by saying that the uranium toning process lends itself most admirably to pictures produced upon Fry's Roughest Paper (Naturalistic Brand), some of my efforts on that make of paper, afterwards toned with uranium exactly in the manner herein described, having met with kindly recognition from exhibition judges.

CHAPTER X.

I SUPPOSE no subject connected with photography
creates more alarm in the mind of the amateur than
enlargement calculations, and yet, despite this, nothing is
simpler.   It involves a very little arithmetic of the most
elementary kind, which saves the would be enlarger no end
of trouble, and yet it is generally shirked because of the
fancied difficulties.

I do not intend to say much on this subject, as I wish to
rely upon the simplicity of my illustrations to convince the
amateur that in calling his difficulties " fancied " I have
not exaggerated their unimportance.

Suppose it is required to enlarge a quarter-plate negative
to five times its size, and a 6-inch focus lens is in use, how
far should the negative be placed from the lens, and the
lens from the screen ?   This is how to find out.   Add 1 to
the number of times you are going to enlarge, which is
five ; five and one == 6.   Now multiply that figure by the
focus of the lens=6.   Thus, 6 × 6 = 36.   Thirty-six inches
is the distance the lens must be from the screen   To find
out how far the negative must be from the lens, divide the
above figure (36) by the number of times of enlargement,
and the answer will be the distance required to be known.
Thus, $\frac{36}{5}=7\frac{1}{5}$.   Could anything be simpler ?

But there is another way, and a simpler one, if possible,
which will save the trouble of any calculations.   All that is
required is to look at the head of the columns of the accompanying table for the number of times you want to
enlarge your picture (let us suppose it to be 6) ; and in
the left hand column for the focus of the lens (which we
will suppose to be 7), and then in the square opposite the
6 and under the 7 you will see the figures 49 and 8¼.   The

first tells how far the screen must be from the lens, the second how far the negative must be from the lens.

REVISED TABLE FOR ENLARGEMENTS.

Focus of Lens, inches.	TIMES OF ENLARGEMENT AND REDUCTION.							
	1 inches.	2 inches.	3 inches.	4 inches.	5 inches.	6 inches.	7 inches.	8 inches.
2	4 4	6 3	8 2⅔	10 2½	12 2⅖	14 2⅓	16 2 2/7	18 2¼
2½	5 5	7½ 3¾	10 3⅓	12½ 3⅛	15 3	17½ 2 11/12	20 2 6/7	22½ 2 13/16
3	6 6	9 4½	12 4	15 3¾	18 3⅗	21 3½	24 3 3/7	27 3⅜
3½	7 7	10½ 5¼	14 4⅔	17½ 4⅜	21 4⅕	24½ 4 1/12	28 4	31½ 3 15/16
4	8 8	12 6	16 5⅓	20 5	24 4⅘	28 4⅔	32 4 4/7	36 4½
4½	9 9	13½ 6¾	18 6	22½ 5⅝	27 5⅖	31½ 5¼	36 5 1/7	40½ 5 1/16
5	10 10	15 7½	20 6⅔	25 6¼	30 6	35 5⅚	40 5 5/7	45 5⅝
5½	11 11	16½ 8¼	22 7⅓	27½ 6⅞	33 6⅗	38½ 6 5/12	44 6 2/7	49½ 6 3/16
6	12 12	18 9	24 8	30 7½	36 7⅕	42 7	48 6 6/7	54 6¾
7	14 14	21 10½	28 9⅓	35 8¾	42 8⅖	49 8⅙	56 8	63 7⅞
8	16 16	24 12	32 10⅔	40 10	48 9⅗	56 9⅓	64 9 1/7	72 9
9	18 18	27 13½	36 12	45 11¼	54 10⅘	63 10½	72 10 2/7	81 10⅛

CHAPTER XI.

NOT the least important part of a print or enlargement
is its *final complete appearance*, and I therefore add a
few words upon mounting and framing. Only hints, however, can be given, as every one's taste or fancy must
dictate the exact means taken to give the print some
stability or protection. Every photograph that is worth
keeping at all, deserves as much protection as is generally
given to a work of art. If the picture is to be framed, let
the glass be fastened air-tight into the rebate of the frame
by pasting a narrow strip of paper and laying it down to
cover the joint of the glass and the rebate of the frame.
This keeps the air and dust from penetrating to the print.
If a piece of paper be pasted right over the back of
the frame, the whole will then be practically air-tight, and
the life of the print prolonged.

As far as mounting is concerned there is the choice between a cut-out mount and what is generally known as
India Tint or plate marked mounts : the latter being in
imitation of the effect of the copper-plate mark impressed
upon the paper in the printing of an engraving. To make
these oneself is not possible, but a very artistic mounting
board can be easily made to suit and fit each and every
print. Such modifications as are necessary to meet varying
circumstances will suggest themselves to the reader, I am
sure.

Suppose it is desired to mount a 12 × 10 print, a piece
of white or tinted card board about 20 × 18 is required,
also a sheet of India or plate paper. This can be obtained
of varying tints in browns and yellows of wholesale
stationers. Trim the print to such a size as suits its composition best. Then cut the plate paper so as to allow a
margin of suitable width all round the print. What this

width is to be, will be determined by the taste of the
worker, but for a 12 × 10 print, from $1\frac{1}{4}$ to $1\frac{3}{4}$ inches will
be nearly right. It is an improvement to allow a little
deeper margin at the bottom or foreground edge of the
picture. Be careful to trim both the picture and the plate
paper with the corners quite rectangular. Lay the India
paper down upon a clean piece of glass. If this is not
available, an enamelled tea-tray borrowed for the occasion
from the household authorities will be a good substitute.
With a brush, rub starch paste over the paper, and lay it
down in the centre of the piece of cardboard. Now take
a piece of soft blotting paper, and with a silk handkerchief
or a roller squeegee, rub the paper down into close contact
with the cardboard. Should there be air-bells they may be
pricked with a needle to let the air out, and then rubbed
down. Now starch the back of the Bromide print as
before described, and place it centrally upon the India
paper. This must also be rubbed down, any bubbles being
pricked. The hole where the needle enters cannot
be seen when dry. Nothing is better than ordinary Glen-
field starch for mounting purposes (unless the cardboard
be very thin indeed). The best way to prepare the starch is
as follows:—In a basin put a tablespoonful of starch. To
it add a teaspoonful of water and rub the starch into a
smooth paste absolutely free from lumps, and of such a
consistence as to be only just liquid or moist. If the
quantity of water is insufficient, more must be added, but a
few drops only at each time, and the whole must be carefully rubbed in to a very stiff but smooth paste. Now pour
actually boiling water upon the paste (which must be stirred
the whole time) until the opaque light blue turns to a deep
transparent blue color. Keep this stirring for five minutes
to avoid lumpiness, add two pieces of loaf sugar, stir until
dissolved, and then allow the starch to cool down. The
sugar will improve the adhesive quality of the starch and
make it keep sweet and fit for use longer. When quite
cold remove the skin which will have formed upon the top
and the starch is fit for mounting photographs with.

When mounted, the prints may have a tendency to
cockle. If the reader has not a large rolling press, and it

is unlikely, he has a very good substitute in the flat iron *genus domesticus*. Use one with a smooth polished face and just so hot as to be uncomfortably warm to the hand. Pass it over the dried mounted print, taking the precaution to interpose a piece of glazed white paper. The texture of the print is very much improved by this method, which may also be used for flattening unmounted bromide prints. In this latter case it is a good plan to soak the print in a solution containing 5% of glycerine, and allow it to dry again before ironing.

If there are any dirty marks upon the mounted print, the best way to remove them is to rub them with a little fine pumice powder or ink-eraser until the marks are gone. Should this move the surface of the film it may be moistened again with a sponge, allowed to dry, and then ironed.

With this chapter my labor—a labor of pleasure—is finished. I hope that my descriptive writing will be found clear and the information given practical and workable. This is not intended as an exhaustive treatise. I trust it will not be found an exhausting one.

FINIS.

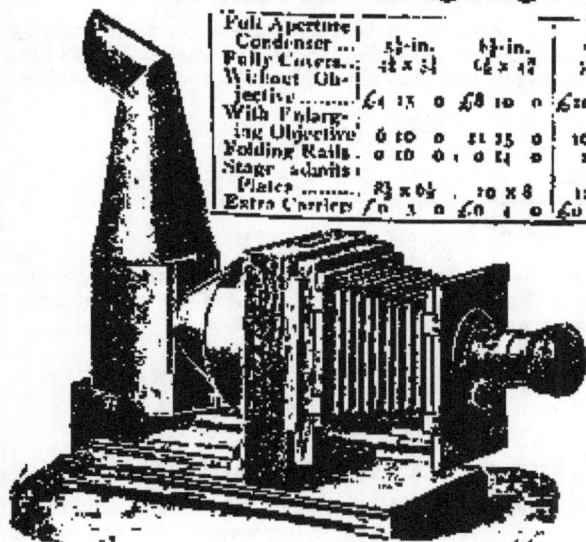

CROUCH'S

LENSES,

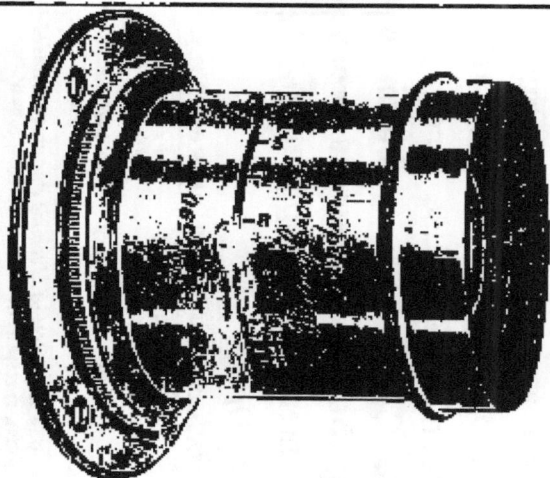

In Aluminium Settings.

Crouch's 'Dresser'

Hand Camera.

Crouch's 'Magazine'

Hand Camera.

Fully Illustrated Catalogues just published. Post Free on application.

HENRY CROUCH, Ld.

141, Oxford St. W.
Optical Works:
66, Barbican, E.C.
London.

FRY'S BROMIDE OPALS.

Our Bromide Opals are sold at the following reduced and popular prices. One Quality—the Best. One grade of Opal only—the Whitest; the Matt surface specially selected. Any person able to use Bromide Paper can make pictures on Fry's Bromide Opals. Every batch is thoroughly tested. Our enlargers are using them by the gross every week. Terms to Dealers and Professional users upon application.

PRICES.—Fry's Matt surface Bromide Opals.

SOLD IN DOZEN PACKETS ONLY.				HALF-DOZEN PACKETS ONLY.		QUARTER DOZEN PACKETS ONLY.		
	3¼ x 3¼ 1/8	4¼ x 3¼ 1/6	6½ x 4⅞ 3/6 per doz.	5½ x 6⅜ 4/-	10 x 8 6/-	12 x 10 9/- per ½-doz.	15 x 12 8/6	20 x 16 12/-	18 x 24 inches. 18/- per ½-doz.
Or free by parcel post, including packing, for ⅓ doz. ask with order only.	1/8	1/9	4/3	1/11 4/8	7/6	20/6	8/-	Too heavy.	Too heavy.

Packed for post with utmost care, but at customer's risk. *Instructions for use, and test pieces of Bromide Paper are included in every package.* To buyers of our Opals a practical demonstration is given, if required, at our London Show and Demonstration Rooms, 5, Chandos Street, Charing Cross.

THE FRY MANUFACTURING CO., London and Kingston-on-Thames.